United States Government Accountability Office

Report to Congressional Committees

I0463616

November 2016

COMMERCIAL SPACE

FAA Should Examine How to Appropriately Regulate Space Support Vehicles

November 2016

COMMERCIAL SPACE

FAA Should Examine How to Appropriately Regulate Space Support Vehicles

Highlights of GAO-17-100, a report to congressional committees

Why GAO Did This Study

As the commercial space transportation industry has grown significantly in the last decade, a related industry has emerged that plans to complement the commercial space industry by using vehicles called space support vehicles to conduct space-related activities, but not launch into space.

The U.S. Commercial Space Launch Competitiveness Act of 2015 includes a provision for GAO to review the uses for space support vehicles and services and any barriers to their use. This report addresses stakeholder views on (1) potential uses for space support vehicles, (2) challenges that companies may face when attempting to use these vehicles, and (3) how these vehicles should be regulated. GAO reviewed prior GAO and industry reports, relevant laws and regulations, and interviewed officials on two proposals for regulating space support vehicles. GAO interviewed officials at FAA and the National Aeronautics and Space Administration and 37 legal experts and stakeholders from industry organizations, launch companies, space support companies, and spaceports—identified by agency and industry officials.

What GAO Recommends

The Secretary of the Department of Transportation (DOT) should direct the FAA Administrator to fully examine and document whether the FAA's current regulatory framework is appropriate for space support vehicles and, if not, suggest legislative or regulatory changes, or both, as applicable. DOT provided technical comments; however, it did not comment on the recommendation at this time.

View GAO-17-100. For more information, contact Gerald L. Dillingham, Ph.D. at (202) 512-2834 or dillinghamg@gao.gov

What GAO Found

Company officials GAO interviewed identified potential uses for "space support vehicles"—which include a variety of aircraft from high-performance jets to balloons and the aircraft portion of a hybrid launch systems (a vehicle that contains elements of both an aircraft and a rocket-powered launch vehicle)—but the size of the market for these uses is unclear. Company officials said they plan to use space support vehicles to train spaceflight participants and to conduct research in reduced gravity environments. For example, some company officials said they would like to use high-performance jets to train future spaceflight participants by exposing them to physiological and psychological effects encountered in spaceflight. Other company officials said they would like to use space support vehicles to research how objects or people react in reduced gravity environments. It is difficult to know the size of the market for spaceflight training and research as GAO found no studies on these markets. However, stakeholders said they expect interest in research to increase.

Some company officials said the Federal Aviation Administration's (FAA) regulatory framework presents a market challenge because companies cannot get FAA approval to use the aircraft they would like to use to carry passengers or cargo for compensation, thus limiting their ability to operate in the market. FAA's Office of Aviation Safety (AVS) regulates aircraft that companies would like to use as space support vehicles by issuing standard and experimental certificates that help ensure safety. While officials from two companies GAO interviewed have received standard aircraft certification for their space support vehicle, others said the standard certification process is lengthy and not designed for the type of vehicles they would like to use, such as unique, single-production aircraft or retired military jets. In addition, FAA regulations do not allow companies to receive compensation for carrying people or property on an aircraft operating under an experimental certificate. As a result, some of the companies we interviewed have training operations in other countries where they can receive payment for the activity. Further, FAA's Office of Commercial Space Transportation (AST)—the office that regulates commercial space activities—is only authorized to regulate commercial space activities, such as launches, focusing on the safety of third parties. According to FAA officials, a statutory or regulatory change would be needed to allow companies to use space support vehicles that do not meet AVS's standard certification requirements for compensation.

Stakeholders GAO interviewed have mixed views on how FAA should regulate space support vehicles; some companies believe the current regulatory approach is appropriate, while others believe the system should be changed in the face of new technology and commercial space development. While FAA has taken steps to assess the licensing and permitting process for hybrid launch vehicles, it has not assessed whether space support vehicles are needed and if it should propose changes that would accommodate all aircraft that could be used as space support vehicles. Thus, some U.S. company officials said they are delaying investments in space support vehicles, and therefore, it is uncertain if they will be able to use them to meet the future needs of the commercial space transportation industry.

_____ United States Government Accountability Office

Contents

Abbreviations

AST	Office of Commercial Space Transportation
AVS	Office of Aviation Safety
COMSTAC	Commercial Space Transportation Advisory Committee
CSF	Commercial Spaceflight Federation
FAA	Federal Aviation Administration
FTE	full-time equivalent
LODA	Letter of Deviation Authority
NASA	National Aeronautics and Space Administration

U.S. GOVERNMENT ACCOUNTABILITY OFFICE

441 G St. N.W.
Washington, DC 20548

November 25, 2016

The Honorable John Thune
Chairman
The Honorable Bill Nelson
Ranking Member
Committee on Commerce, Science, and Transportation
United States Senate

The Honorable Lamar Smith
Chairman
The Honorable Eddie Bernice Johnson
Ranking Member
Committee on Science, Space, and Technology
House of Representatives

The U.S. commercial space industry has seen significant development in the past decade, generating hundreds of millions of dollars in revenue by launching satellites and other payloads[1] into space while working toward developing space tourism. More recently, a related industry has emerged that is planning to complement the commercial space industry primarily by using a variety of aircraft, from high-performance jets to balloons and hybrid launch vehicles,[2] to conduct space-related activities but not launch into space.[3] For example, these companies plan to use aircraft to test equipment bound for the international space station, conduct microgravity research, and train future space flight participants[4] (or tourists) hoping to ride in future commercial launch vehicles.[5]

---

[1] Payloads are space probes, on-orbit vehicles, or spacecraft that can carry humans, animals, or cargo.

[2] Such vehicles contain elements of both aircraft and rocket powered vehicles.

[3] Some companies would like to conduct space support activities through ground based means. These will be discussed later in the report.

[4] Federal law currently defines a "space flight participant" as "an individual, who is not crew or a government astronaut, carried within a launch vehicle or reentry vehicle." 51 U.S.C. § 50902 (20).

[5] Some participation by such "space tourists" has occurred in other countries.

Due to an interest in identifying the best way to regulate and at the same time develop a space support industry, the U.S. Commercial Space Launch Competitiveness Act[6] includes a provision for us to look at the use of space support services and vehicles and any barriers to their use.[7] There is, however, neither a regulatory nor a statutory definition nor an agreed-upon industry definition of space support services or space support vehicles. We use the term "space support vehicles" to refer to aircraft used in support of commercial space activities, but not as part of a launch.[8] This report discusses stakeholders' views on:

- potential uses for space support vehicles in the commercial space industry,

- challenges companies face as they attempt to use space support vehicles, and

- approaches for regulating space support vehicles.

To address all of the objectives, we interviewed commercial space-industry stakeholders, including officials from the Federal Aviation Administration's (FAA)[9] Offices of Commercial Space Transportation and Aviation Safety and the Congressional Research Service as well as 37 commercial space-industry stakeholders, including representatives of 13 launch companies, 9 space support service providers, and 5 spaceports.[10] We attended a Commercial Space Transportation Advisory Committee (COMSTAC) meeting and interviewed industry organizations

---

[6] Pub. L. No. 114-90, § 116, 129 Stat. 704, 717 (2015).

[7] For the purpose of this report and after conversations with stakeholders in industry and within the Federal Aviation Administration (FAA)—the agency that regulates commercial space transportation and civil aviation—we are grouping these disparate activities under the heading of the "commercial space-support industry." This term encompasses activities: (1) currently undertaken by the private sector, (2) in the planning and development stages, and (3) private industry would like to undertake but are currently prohibited from conducting.

[8] Launch activities are covered under 51 U.S.C. Chapter 509.

[9] The Federal Aviation Administration is within the Department of Transportation.

[10] Spaceports are sites used for commercial space launches. Space support companies include companies that would like to use aircraft for space support activities such as training and carrying scientific payloads.

including the Commercial Spaceflight Federation[11] and the Aerospace Industries Association. We also conducted a site visit to Florida, which we selected based on the location of commercial space companies, and met with officials from Space Florida,[12] a spaceport, and a space support-service provider. We identified stakeholders by asking for a list of potential stakeholders from FAA's Office of Commercial Space Transportation and the Commercial Spaceflight Federation as well by identifying stakeholders from our previous work on commercial space. We contacted these stakeholders and identified additional stakeholders through a snowball technique in which stakeholders identified additional contacts during interviews. We approached all companies and individuals on the list and scheduled interviews with interviewees who responded.[13] The results of these interviews are non-generalizable. For a full list of stakeholders we interviewed, see appendix I.

To identify the commercial space industry's potential uses for space support vehicles,[14] we also reviewed prior GAO reports on commercial space issues, FAA's Annual Compendium of Commercial Space Transportation, a National Research Council report requested by the National Aeronautics and Space Administration (NASA) related to astronaut training,[15] and industry reports on the commercial space market. We interviewed officials from NASA to learn about astronauts' training techniques that would inform spaceflight participants' training techniques. To identify challenges that companies face when attempting to use space support vehicles, we reviewed relevant laws and regulations and interviewed three individuals who teach at institutions of higher education and whom we interviewed through an Internet search on

[11] The Commercial Spaceflight Federation is an organization that represents the commercial space industry. There are two levels of membership. Executive Members include commercial spaceflight developers, operators, and spaceports. Associate Members include suppliers supporting commercial spaceflight, with recent members including suppliers of mission support services and suppliers of training, medical and life-support products and services.

[12] Space Florida is an aerospace economic development agency for the state of Florida.

[13] In one case, a stakeholder responded in writing to questions and a formal interview was not conducted. In addition, 14 did not respond or declined our request for an interview.

[14] When we asked companies about their plans for space support vehicles, some had business plans, but these were proprietary and not shared with us.

[15] National Research Council, *Preparing for the High Frontier: The Role and Training of NASA Astronauts in the Post-Space Shuttle Era,* (Washington, D.C.: 2011).

academic space-law programs and from prior work to obtain perspectives on the legal issues involved with space support vehicles. To obtain stakeholder views on how space support vehicles should be regulated, we interviewed FAA officials on their developing proposal on the regulation of hybrid launch vehicles used in non-launch and non-reentry activities and Commercial Space Federation officials on their thoughts on the regulation of space support vehicles. We also reviewed the Standards for Internal Control in the Federal Government[16] to identify how internal control systems should address changing conditions and compared pertinent standards with FAA's efforts to adapt its regulations to changing technology and business plans in the commercial space support sector.

We conducted this performance audit from February 2016 through November 2016 in accordance with generally accepted government auditing standards. Those standards require that we plan and perform the audit to obtain sufficient, appropriate evidence to provide a reasonable basis for our findings and conclusions based on our audit objectives. We believe that the evidence obtained provides a reasonable basis for our findings and conclusions based on our audit objectives.

Background

Although little is known about the emerging space support vehicle industry, the U.S. commercial space-launch industry generated $617 million in revenue in 2015 and has experienced significant growth in the past half-decade. FAA reported that its licensed launches have increased 60 percent and industry revenue has increased 471 percent since 2012. We've previously reported that the industry has experienced growth in the number and complexity of launches and growth in demand for space launches.[17] Furthermore, the industry is developing new types of reusable launch vehicles, which could reduce launch costs.[18]

Components of the U.S. commercial space transportation industry include:

---

[16] *Standards for Internal Control in the Federal Government,* GAO-14-704G (Washington, D.C.: Sept. 10, 2014).

[17] GAO, *Commercial Space: Industry Developments and FAA Challenges,* GAO-16-765T (Washington, D.C.: June 22, 2016).

[18] GAO, *Federal Aviation Administration: Commercial Space Launch Industry Developments Present Multiple Challenges,* GAO-15-706 (Washington, D.C.: Aug. 25, 2015).

- *Launch Companies:* These companies launch satellites or other payloads into space. Their clients include governments, other companies, and individuals. Currently, the industry is launching non-human payloads using rockets and has yet to launch space flight participants. At the same time, some launch companies are developing hybrid launch systems, which contain elements of both aircraft and rocket-powered vehicles.[19] Companies we interviewed plan to use these vehicles—which take off as an aircraft and then launch the spacecraft once reaching a certain altitude—to launch non-human payloads and to transport spaceflight participants to space.[20] For example, WhiteKnightTwo plans to carry Virgin Galactic's SpaceShipTwo aircraft to an altitude of 50,000 feet where it will air launch SpaceShipTwo and the participants on board into space. According to representatives of launch companies we interviewed, while most of launch companies' current activities are focused on launches, they may also use the aircraft component of a hybrid launch system for non-launch activities. For example, Virgin Galactic has also considered using WhiteKnightTwo to carry scientific payloads into conditions that would simulate spaceflight.

---

[19] Rockets are aircraft propelled by ejected expanding gases generated in the engine from self-contained propellants and not dependent on the intake of outside substances. It includes any part which becomes separated during the operation.

[20] Some of these companies are also launching payloads in vertical rockets, which are not space support vehicles.

Figure 1: Virgin Galactic's WhiteKnightTwo Carrying SpaceShipTwo

Source: © 2016 Mark Greenberg/Virgin Galactic. | GAO-17-100

- *Spaceports:* Spaceports are FAA-licensed launch or reentry sites used for commercial space launches and reentries that are developed by private companies and/or states. Spaceports can be co-located with federal sites or commercial or general aviation airports. Spaceports generally include launch pads and runways as well as other infrastructure, such as hangar space, and services, such as emergency services to be used by commercial space companies. As of June 2016, there were 10 FAA-licensed nonfederal launch sites.

- *Space Support Companies:* This category includes companies whose business plans focus on training future spaceflight participants. This training simulates conditions encountered in space and can either be accomplished on the ground using pools and centrifuges or in the air using aircraft, hence the interest of companies in the commercial space support vehicle industry. In addition to training, space support companies offer or plan to offer other services, such as carrying scientific payloads for micro gravity experiments and repositioning cargo from one location to another through the air.

FAA's Office of Aviation Safety (AVS) oversees civil aviation activities in the United States. Thus, it regulates aircraft that may be used for space support activities. AVS is responsible for certifying the airworthiness of aircraft,[21] pilots, mechanics, and others whose work affects the safety of those aircraft. When certifying aircraft, AVS inspectors review aircraft engines, propellers, parts, and equipment, including avionics, to provide a reasonable expectation of safety. In addition, AVS is responsible for certifying all operational and maintenance enterprises in domestic civil aviation. In fiscal year 2016, AVS had 7,246 full-time equivalent (FTE) employees and a budget of $1.26 billion.

The Office of Commercial Space Transportation (AST) is the office within the FAA responsible for overseeing and coordinating the conduct of commercial launch and reentry operations and issuing and transferring licenses and permits authorizing such activities.[22] Unlike AVS, AST has a dual mandate to (1) protect the public health and safety (people not participating in the launch, i.e., third parties), the safety of property, and national security and foreign policy interests of the United States during commercial launch and reentry activities and (2) encourage, facilitate, and promote U.S. commercial space transportation.[23] The Commercial Space Launch Amendments Act of 2004[24] instructed the Department of Transportation (DOT) to promote the continuous improvement of the

---

[21] 49 U.S.C. § 44704(d).

[22] The Commercial Space Launch Act of 1984, Pub. L. No. 98-575, 98 Stat. 3055 (1984) designated the Department of Transportation (DOT) as the federal agency responsible for overseeing and coordinating the conduct of commercial launch operations, issuing and transferring commercial launch licenses authorizing such activities, and protecting the public health and safety, safety of property, and national security interests and foreign policy interests of the United States. The DOT delegated these responsibilities to the FAA's Office of Commercial Space Transportation.

[23] 51 U.S.C. § 50903.

[24] Pub. L. No. 108-492, 118 Stat. 3974 (2004).

safety of launch vehicles designed to carry humans.[25] Prior to launch, space flight participants must provide written consent to participate.[26] In fiscal year 2016, AST had 92 FTE employees and a budget of $17.8 million.

Stakeholders Identified Potential Uses for Space Support Vehicles, but the Size of the Market Is Unclear

Industry Stakeholders Discussed Uses for Space Support Vehicles—Spaceflight Participant Training and Research

Spaceflight Participant Training

Stakeholders identified spaceflight participant training as one potential use for space support vehicles. FAA regulations require only minimal training for spaceflight participation. Specifically, operators are required to train each space flight participant how to respond to emergency situations, such as smoke, fire, and the loss of cabin pressure.[27] However, some companies are interested in providing training beyond the minimal requirements to potential space flight participants.

Stakeholders we interviewed disagreed on the best way to train future spaceflight participants. Some industry stakeholders (13 of 37) told us

---

[25] Chapter 509 currently prohibits the FAA from issuing regulations governing the design or operation of a launch vehicle to protect the health and safety of crew and space flight participants until 2023, unless it does so in response to a serious or fatal injury, or an event that posed a high risk of causing a serious or fatal injury to crew, government astronauts, or space flight participants. 51 U.S.C. § 50905(c).

[26] 51 U.S.C. § 50905(b)(5)(C).

[27] 14 C.F.R. § 460.51.

that future space flight participants will need to receive training in high-performance aircraft. Specifically, 4 of 9 space support companies, 3 of 13 launch companies, 2 of 5 spaceports, and 4 of 10 other stakeholders argued that it is necessary for customers to fully understand what they will experience and the only way to replicate this is through training in high-performance jets. Stakeholders explained that factors that can be simulated in high performance jets but not through other means include the stress of a confined environment and exposure to the physiological and psychological effects of spaceflight. Further, four stakeholders said this training is necessary to become acquainted with g-forces[28] involved with spaceflight and that space support vehicles are best able to provide this training.[29] NASA officials also stated that familiarization with and experiencing high-g environments while performing time critical communication is important preparation for spaceflight participants. See figure 2 for examples of space support vehicles, including high-performance jets (on left) and the modified Boeing 727 (on right). In addition, six stakeholders explained that due to the high cost of spaceflight, getting this experience would be important for someone who is considering space tourism. However, as discussed below this training currently is not available in the U.S.

On the other hand, 13 stakeholders report training can be accomplished through currently allowed means, including standard certified aircraft and ground based training. Five of 9 space support companies, 5 of 13 launch companies, and 3 of 10 other stakeholders reported that it is critical to expose future spaceflight participants to the conditions they will encounter in space, but these conditions can be replicated through means other than high-performance jets. Two space support companies said they provide ground-based training through centrifuges, pools, and instruction in space-related topics such as the physiological and psychological effects of space travel. One launch company reported to us that spaceflight participants will only need to know how they will react to microgravity but will not need to know how to accomplish tasks in microgravity. According to some stakeholders, this acclimation to microgravity can be accomplished through parabolic flight, centrifuges,

---

[28] G-forces are a force on a body as a result of acceleration or gravity. One g is equivalent to the force of gravity on Earth's surface.

[29] While this issue was raised by stakeholders in some of our interviews, we did not discuss it in all interviews.

and pools.[30] One company is currently offering a microgravity experience in a Boeing 727 with an interior that has been modified to accommodate passengers to this activity, and another company has considered entering this market with a similar aircraft. Two of 13 launch companies reported they do not anticipate a high training burden for future customers and thus would probably not utilize aircraft training or centrifuges.

The remaining 12 of the 37 stakeholders we interviewed did not comment on this issue.

Figure 2: Examples of Space Support Vehicles

Sources: Starfighters Aerospace (left) and Steve Boxall/Zero Gravity Corp. (right). | GAO-17-100

Further, stakeholders representing spaceports have proposed that spaceports are the proper place to host spaceflight participant training. Spaceport stakeholders we interviewed said that spaceports provide runways, launch pads, hangars, and other services such as emergency response services for commercial space-transportation companies. As mentioned above, representatives of two of the five spaceports we spoke to thought space flight training should be provided in high-performance jets. Two stakeholders we interviewed at spaceports also expressed

---

[30] Zero G Corporation, which operates parabolic flights, describes the flights as "[b]efore starting a parabola, G-FORCE ONE flies level to the horizon at an altitude of 24,000 feet. The pilots then begins to pull up, gradually increasing the angle of the aircraft to about 45° to the horizon reaching an altitude of 32,000 feet. During this pull-up, passengers will feel the pull of 1.8 Gs. Next the plane is 'pushed over' to create the zero gravity segment of the parabola. For the next 20-30 seconds everything in the plane is weightless."

interest in hosting companies that provide spaceflight participant training and see it as a potential source of new revenue, either now or when the tourism industry evolves and sends customers into space. One spaceport operator we interviewed sees spaceflight participant training as economically beneficial for the communities surrounding spaceports.

Research on the Effects of Zero Gravity

In addition to spaceflight participant training, some stakeholders we spoke to identified microgravity research as a potential use for space support vehicles. Microgravity research uses reduced gravity to understand how objects or people will react in reduced gravity environments (such as orbit). Microgravity can be provided through parabolic flights. NASA officials told us microgravity flights are used to test equipment that will be sent into orbit, including for example exercise equipment and 3D printers. According to FAA, one company currently conducts microgravity research using a Boeing 727 with a standard airworthiness certificate. Other companies have proposed using retired military aircraft to fly scientific payloads for researchers.[31]

Data on the Size of the Space Support Market Is Virtually Non Existent

Spaceflight Participant Training

It is difficult to determine the size of the market for the use of space support vehicles for training because we have found no publicly available studies on the size of the spaceflight participant training market, and companies we interviewed told us they have not conducted their own market analysis. However, companies within the industry provided a wide range of estimates of the size of a potential training market. Estimates of the training market reported by stakeholders are often dependent on the size of the overall space tourism market and the training burden launch companies anticipate for their customers. One industry study that found that there are around 8,000 individuals with the money and inclination to take a space tourism flight by 2022.[32] However, it's not clear how many of these 8,000 individuals would choose to purchase training from training

---

[31] In addition to the above uses, individual stakeholders proposed a variety of other potential uses for space support vehicles including advertising and marketing, repositioning of objects such as rockets and other vehicles, and as a drop vehicle for testing other aircraft.

[32] Tauri Group, *Suborbital Reusable Vehicles: A 10-Year Forecast of Market Demand,* 2012.

providers, and the number would likely depend on whether launch companies require training for passengers, how intensive the training will be, and whether they will contract for this training or offer it in house. Further, studies of the tourism market that we identified are dated and may not reflect current industry conditions.

Research

According to stakeholders we interviewed, there are no studies available on the research market; however, they said that research is a growing segment of the space support market. One stakeholder reported that the research market is the most robust commercial space market that currently exists. It is unclear how many aircraft operators are currently supplying aircraft services for research, but five stakeholders we interviewed expressed interest in using their aircraft to carry scientific payloads for researchers. Based on our interviews, the main customers for this service include universities, the government, and private sector organizations.

FAA's Current Statutory and Regulatory Framework Limits Market Development

FAA's Standard Aircraft Certification Process Is Intended to Ensure Passenger Safety; However, Companies Said They Face Challenges Because the Process is Lengthy and Costly

Two of the companies that we interviewed obtained standard aircraft certification from FAA for aircraft that could be classified as space support vehicles. However, as representatives from one company explained, the certification process was lengthy and expensive.[33] One of the companies received certification to operate parabolic flights using a retrofitted Boeing 727. These flights provide a weightlessness experience that could be used for spaceflight participant training (see fig 3). A company representative told us that the certification process took 18 months and cost millions of dollars. The other company uses a certified aircraft to move its rocket that would be used in a launch from one place to another. While two companies were able to obtain standard aircraft certification for what could be considered space support vehicles, other stakeholders said that the aircraft certification process may not be economically feasible for companies due to the cost of meeting the requirements. For example, representatives of one company said they have considered acquiring a high-performance jet for spaceflight participant training, but that the market for spaceflight participant training would not support the investment needed to purchase the aircraft and go through the current AVS certification process.

---

[33] In 2010, we found that while the certification process generally worked well, some industry stakeholders have had negative experiences that have led to costly delays. For example, one aviation representative said that his company incurred a delay over 5 years and lost millions of dollars when attempting to obtain FAA certification. GAO, *Aviation Safety: Certification and Approval Processes Are Generally Viewed as Working Well, but Better Evaluative Information Needed to Improve Efficiency*, GAO-11-14 (Washington, D.C.: Oct. 7, 2010).

Figure 3: Example of an Aircraft Holding a Standard Certificate Being Used for Parabolic Flight

Source: Al Powers/Zero Gravity Corp. | GAO-17-100

In addition, FAA's standard aircraft-certification process is not well suited for the types of aircraft that space support companies would like to use. As mentioned earlier, AVS regulates the safety of aircraft by certifying aircraft to provide a reasonable expectation of safety. According to FAA officials, aircraft manufacturers are typically set up to work with FAA on the certification process, which is an on-going process as the aircraft are designed and built. The process is not designed for single-production aircraft like those launch companies are developing or retired military jets that companies would like to use for spaceflight participant training.[34] Further, if an aircraft with a standard airworthiness certificate is modified or used for another purpose than its original purpose, including for space support, FAA regulations for the standard aircraft certification process require documentation of all modifications that demonstrates that these modifications comply with applicable regulations.

---

[34] FAA's *Advisory Circular 21-13* established a process for certifying military aircraft for civilian use. However, according to an FAA official, this process applies to aircraft, such as cargo planes, that have both a military and civil version and not the type of aircraft that companies would like to use as space support vehicles.

Companies Are Prohibited from Receiving Compensation under Experimental Certificates

AVS allows certain aircraft to fly under an experimental certificate, but companies are prohibited from operating these aircraft for carrying persons or property for compensation and hire—meaning companies cannot receive money for carrying passengers or cargo.[35] Operators can apply for experimental certificates for unique aircraft that have not been approved under the AVS certification process. Experimental certificates can be issued for:

- research and development;

- showing compliance with regulations;

- crew training;

- exhibition (such as air shows or movie production);

- air racing; and

- conducting market surveys.

FAA has provided experimental certificates for some vehicles that could be used for space support services. For example, experimental certificates have been issued for aircraft that are part of a hybrid launch vehicle system for testing and further aircraft development. However, because current regulations do not allow the owners of these experimentally certified vehicles to carry persons or property for compensation, companies are not allowed to use experimentally certified aircraft for space flight participant training or to transport cargo on a hybrid launch system.

The restriction on using experimentally certified aircraft to carry persons or property for compensation has limited some companies' ability to operate in the space-support services market. Three stakeholders we interviewed said they would like to operate space support vehicles, but are having a difficult time securing funding from investors because of market uncertainty and not knowing if they will be allowed to operate them. In addition, some of the companies we interviewed have training operations in other countries because they are not allowed to operate specific aircraft in the United States under current laws and regulations. Further, one company we interviewed has a spaceflight-participant training program that received a safety approval from AST but does not own a vehicle for the training. Company representatives said that they do

---

[35] 14 C.F.R. §91.319(a)(2).

not want to invest in a space support vehicle until they know if they will be allowed to operate it.

Allowing companies to use experimentally certified aircraft for compensation or hire would require a regulatory change. Some of the stakeholders that we interviewed said that a Letter of Deviation Authority (LODA) from an experimental certification may be an option to be able to operate for compensation or hire, but FAA officials we interviewed said that LODAs only apply to pilot training and cannot be used for spaceflight participant training.

Experimental Permits Only Apply to Launches

As described previously, AST issues licenses and permits for commercial space launches. This process includes issuing experimental permits for the development of hybrid vehicles that are connected with a launch activity. When the aircraft component of a hybrid launch vehicle is used in non-launch operations, companies must go through AVS's certification process to obtain an experimental certificate and operate under aviation regulations. When hybrid launch vehicles are being developed for launch activities, companies operate under commercial space regulations, and FAA may issue an experimental permit or a license. Similar to experimental certificates, companies with experimental permits are prohibited from carrying property or human beings for compensation and hire, according to statute. While companies can perform activities, such as conducting test flights, they may not use hybrid launch vehicles to receive compensation for carrying persons or property. For example, companies would not be able to receive compensation for carrying a researcher. When issuing experimental permits, AST's process focuses on minimizing risks to ensure the safety of third parties. As discussed below, FAA is developing a report that would help address this issue specifically for hybrid launch vehicles used in non-launch non-reentry operations.

Stakeholders Have Mixed Views on the Appropriate Approach for Regulating Space Support Vehicles; FAA Has Taken Some Steps to Assess Its Regulatory Framework

Some Options for Regulating Space Support Vehicles Would Require Statutory or Regulatory Changes

Through our discussions with FAA and interviews with stakeholders, we identified potential options for regulating space support vehicles (see table 1). One of these options is to keep in place the current process using standard airworthiness certificates for regulating aircraft that companies would like to use for space support vehicles. Other options would require statutory and/or regulatory changes to allow for the operation of space support vehicles. Each of these regulatory options raises issues to be considered.

Table 1: Options for Regulating Space Support Vehicles and Related Considerations

Process for Regulating Space Support Vehicles	Federal Aviation Administration Office	Statutory and Regulatory Change Required	Considerations
Keep current process of issuing standard certificates for aircraft	Aviation Safety (AVS)	No statutory or regulatory change needed	• Process is intended to provide a reasonable expectation of safety for the aircraft. • Some companies currently provide space support services under this process. • Process is not designed for uniquely produced aircraft. • Military aircraft that companies would like to use for space support services are unlikely to be approved.

Process for Regulating Space Support Vehicles	Federal Aviation Administration Office	Statutory and Regulatory Change Required	Considerations
Revise experimental certificate for aircraft to allow companies to receive compensation for space support services	AVS	• Regulatory change needed to allow for compensation • Statutory or regulatory change needed to define what space support services or activities would be allowed	• Need to determine if providing a reasonable expectation of safety for the aircraft would still be applicable. • Regulatory change allowing for compensation may apply to activities other than support activities unless the change is limited to those activities.
Make AST responsible for oversight of non-launch flight of space support vehicles related to commercial space transportation	Commercial Space Transportation (AST)	• Statutory and regulatory changes needed • To make AST responsible for overseeing space support vehicles • To allow for compensation • To define services allowed, i.e., how to determine if an activity is related to commercial space transportation • To define the circumstances under which these services can be allowed	• May increase AST's workload. • May focus on hybrid vehicles. Other aircraft may not be covered. • Could streamline regulatory process for some companies because they would be dealing with one entity within FAA. • Could change the safety expectations to an informed consent regime as is currently in effect for commercial space transportation.[a]

Source: GAO analysis. | GAO-17-100.

[a]51 U.S.C. § 50905(b)(5)(A)-(C) requires that spaceflight participants be informed of the risks involved, and 51 U.S.C. § 50914(b) requires spaceflight participants, among other things, to sign a reciprocal waiver of claims (also called a cross waiver) with the federal government—which means that the spaceflight participant agrees not to seek claims against the federal government if an accident occurs.

Options and Related Considerations for Regulating Space Support Vehicles

When asked what changes, if any, should be made to FAA's current regulatory process to oversee aircraft that might provide support services for the commercial space transportation industry, stakeholders had mixed views. Twenty-five of the 37 stakeholders expressed some opinion about changing the regulatory process. While 11 of these 25 stakeholders did not see a problem with the current approach for space support vehicles, 14 of 25 expressed an interest in a change.

Eleven of the 25 stakeholders who expressed an opinion said that the current regulatory process under AVS is the best approach for regulating space support vehicles. These stakeholders prefer the currently regulatory approach for the following reasons:

- They said that the AVS certification process could best protect participants and third parties from potential safety risks. They said that the technical expertise for ensuring that aircraft are safe for participants is within AVS.

- Some of the proposed space support activities—such as using retired military jets to provide spaceflight participant training—are not legitimate space activities but are in fact recreational aviation activities and should therefore be regulated along with other aviation activities.

- Some stakeholders said that the current AVS system is preferable because AST is overburdened, and its staff needs to focus on other commercial space activities, such as issuing launch licenses.[36]

- Some stakeholders would like these activities to remain under AVS to keep them separate from commercial space launch activities, especially in the public eye. For example, three stakeholders we interviewed were worried that a crash involving a space support vehicle might negatively impact the entire commercial space transportation industry. Two stakeholders expressed concern that moving space support vehicles to AST would place these activities under the informed consent regime.[37] Thus, accidents might impact the safety numbers[38] and mar the image and thus of attractiveness of the future commercial space tourism industry.

Other stakeholders we interviewed believe that AST would be a preferable FAA office to provide regulatory oversight to AVS. Specifically, 6 of the 25 stakeholders that expressed an opinion said that space support vehicles should be regulated under AST, citing the following reasons:

- Two stakeholders interviewed said that AST staff are familiar with the commercial space transportation industry and are therefore in the best

---

[36] In 2015, we found that FAA requested additional staff to keep pace with the rapid growth of the U.S. commercial space transportation industry. We also found that FAA did not have certain workload metrics regarding its oversight activities, and thus recommended that FAA provide more detailed information in its budget submissions on its workload. GAO-15-706. In a 2016 testimony statement, we noted that FAA was continuing to develop this information; however, the recommendation has not yet been fully addressed. GAO-16-765T.

[37] 51 U.S.C. §§ 50905(b)(5)(A)-(C) and 50914 require that spaceflight participants be informed of the risks involved and sign a reciprocal waiver of claims (also called a cross waiver) with the federal government—which means that the party agrees not to seek claims against the federal government if an accident occurs.

[38] According to FAA officials, FAA has recently issued draft guidance that says that these accident numbers would only reflect launch activities.

position to determine if a certain vehicle is necessary for commercial space transportation activities.

- Five stakeholders interviewed said that they are already working with AST for other purposes, such as obtaining launch licenses, and would prefer to continue working with one office to streamline the process.

- Two of the stakeholders who preferred AST cited the office's statutory informed-consent regime, which instead of prohibiting certain commercial space transportation activities, ensures that participants are made aware of the activity's potential risks.

In addition, some stakeholders were interested in a combined approach. Specifically, 6 of the 13 launch companies that we interviewed said that all space support vehicles should be regulated by AVS, except hybrid launch vehicles, which they would prefer to be regulated by AST. See discussion below on FAA's proposal on the regulation of the aircraft portion of a hybrid launch system, when it is operating as a space support vehicle. Two stakeholders said that companies should have the choice to work with AVS or AST.

Further, the Commercial Spaceflight Federation (CSF) has been discussing with its membership how space support vehicles should be regulated. CSF's discussion is focusing primarily on the use of experimental aircraft, such as former military jets, for compensation and hire. CSF representatives indicated that some of its members' view is that Congress should direct the Administrator of FAA to authorize spaceflight training flights through rulemaking, but these flights should remain under AVS. However, they noted that if a company has gone through the current difficult certification process for a certain capability, such as using a standard certified aircraft to provide periods of microgravity through a series of parabolic maneuvers, then companies should not need to use an exemption to replicate this service. Further, to help minimize safety risks if space support vehicle flights were to be authorized, these representatives said that they should begin and end at a spaceport. In addition, passengers should be notified that the aircraft are not certified as safe under AVS's aircraft rules and are not licensed as space transportation—essentially an informed consent regime. According to CSF representatives, a benefit of this approach is that it should enable new capabilities that cannot exist within the current legal and regulatory framework while helping minimize safety risks.

FAA Is Assessing the Licensing and Permitting Process for Some Space Support Vehicles

According to FAA officials, the commercial space transportation industry is evolving and AST and AVS have worked with companies individually to determine how they can legally operate within the current regulatory system. However, FAA officials acknowledge that this issue is potentially growing as more companies try to figure out how to cost-effectively provide what they see as a potential market—supporting commercial space transportation. Federal internal control standards state that conditions affecting an agency, such as FAA, and its environment continually change and that these changing conditions often prompt new risks or changes to existing risks that need to be assessed.[39] FAA has taken steps to assess the licensing and permitting process for hybrid launch vehicles; however, it has not assessed whether space support vehicles are needed to meet the potential research, training, and other needs of the commercial space transportation industry, and if it should propose changes that would accommodate all aircraft that could be used as space support vehicles.

FAA officials said that their views on non-launch and non-reentry operations of hybrid launch vehicles will be expressed in their report that was mandated under the U.S. Commercial Space Launch Competitiveness Act.[40] However, this report only focuses on one type of vehicle—hybrid launch vehicles.[41] As we described previously, hybrid launch systems contain elements of both aircraft and rocket-powered vehicles. These vehicles take off horizontally and then launch the spacecraft once reaching a certain altitude. Currently, the portion of a hybrid launch system that can operate like an aircraft is regulated as

---

[39]GAO-14-704G.

[40] The U.S Commercial Space Launch Competitiveness Act, Pub. L. No 114-90, §105, 129 Stat. 704, 707 (2015). Section 105 required FAA to report within 120 days of the date of enactment on November 25, 2015, on approaches for streamlining the licensing and permitting process of launch vehicles, reentry vehicles, or components of launch or reentry vehicles, to enable non-launch flight operations related to space transportation. According to FAA officials, they did not include other types of vehicles in their proposal because Section 116 of this Act required GAO to gather information about the challenges faced by companies that would like to use space support vehicles, including those other than hybrid launch vehicles used in non-launch, non-reentry operations. According to FAA officials, FAA is waiting to see the results of our report before fully assessing their current regulatory framework.

[41] According to FAA officials, FAA considers a hybrid launch vehicle to be a launch or reentry vehicle that may also be used for non-launch, non-reentry operations. FAA defines hybrid launch system as a launch vehicle, including a suborbital rocket, which may be comprised of two or more components, at least one of which is an aircraft and can operate as an aircraft when not engaged in a launch.

aircraft by AVS through experimental certificates when it is not engaged in a launch.[42]

While the U.S. Commercial Space Launch Competitiveness Act required FAA to report on approaches for streamlining the licensing and permitting process of non-launch, non-reentry operations of hybrid launch vehicles related to space transportation, it did not require FAA to develop a proposed regulatory framework. Although Congress did not require FAA to develop a regulatory framework for vehicles that could be used in support of space activities, federal internal control standards[43] state that federal agencies should identify, analyze, and respond to significant changes that could impact the internal control system—the mechanism by which an entity's oversight provides reasonable assurance that the agency's objectives will be achieved. Further, they state that conditions affecting an agency, such as FAA, and its environment continually change and that these changing conditions often prompt new risks or changes to existing risks that need to be assessed and addressed. While FAA has taken steps to handle the safety issues of these vehicles on an individual basis and is assessing approaches for streamlining the licensing and permitting process of hybrid launch vehicles, it has not examined how its regulatory framework should change, if at all, to address the potential growth and related risks in the use of space support vehicles. Thus, some stakeholders we spoke to are delaying investments in space support vehicles.

Conclusions

The U.S. commercial space-transportation industry has seen significant development in the past decade. As the industry evolves, companies are considering how to provide additional services to support the industry's needs. While some companies are using certified aircraft to provide space support services such as spaceflight participant training, other companies would like to use vehicles such as retired military jets and hybrid launch vehicles to provide such services. FAA's current regulatory framework applies to aircraft; however, the aircraft that some companies would like to use to provide space support services do not fit into this framework. As stakeholders recognized, a change in regulatory regimes may impact safety and streamline the regulatory process. For example, while FAA's current regulations help ensure passenger safety, they also

---

[42] 14 C.F.R. §91.319.

[43] GAO-14-704G.

prevent some companies from providing space support services. While FAA has started considering this issue, especially for hybrid launch systems, it has not determined if space support vehicles are needed to meet the potential research, training, and other needs of the commercial space industry nor fully examined its current regulations as they relate to space support vehicles and determined and documented the results of its assessment. Since FAA has not conducted a comprehensive assessment of how space support activities fit under its aviation or commercial space transportation regulatory regimes, officials from some U.S. companies told us they are delaying investments in space support vehicles. As a result, it is uncertain if companies will be able to use space support vehicles for potentially useful spaceflight participant training and research services to meet the future needs of the commercial space-transportation industry.

Recommendation

To respond to changes in the aviation and commercial space-transportation industries, we recommend that the Secretary of Transportation direct the FAA Administrator to fully examine and document whether the current regulatory framework is appropriate for aircraft that could be considered space support vehicles, and if not, suggest legislation or develop regulatory changes, or both, as applicable.

Agency Comments

We provided a draft of this report to the Department of Transportation for review and comment. DOT is not providing comments on the recommendation at this time, but will provide a detailed response to the recommendation within 60 days of the final report's issuance. DOT provided technical comments, which we incorporated into the report, as appropriate.

In addition, to verify information, we provided a draft of this report to NASA for review and comment. NASA provided technical comments, which we incorporated into the report, as appropriate.

We are sending copies of this report to the Secretary of Transportation, the Administrator of the FAA, and the Administrator of NASA, as well as appropriate congressional committees and other interested parties. In addition this report is available at no charge on the GAO website at http://www.gao.gov.

If you or your staff members have any questions about this report, please contact me at (202) 512-2834 or dillinghamg@gao.gov. Contact points for our Offices of Congressional Relations and Public Affairs may be found on the last page of this report. GAO staff who made key contributions to this report are listed in appendix II.

Gerald L. Dillingham, Ph.D.
Director, Physical Infrastructure Issues

Appendix I: Commercial Space Transportation Industry Stakeholders Interviewed

Table 2: List of Stakeholders Interviewed

Industry category	Organization/individual interviewed
Launch Companies	10 Tanker
	Agile Aerospace
	Blue Origin
	CubeCab
	Firefly Space Systems
	Intuitive Machines
	Orbital ATK
	Sierra Nevada Corporation Space Systems
	SpaceX
	Vulcan Aerospace
	World View Enterprises
	XCOR Aerospace
Space Support Companies	IA Space
	Integrated Spaceflight Services
	International Flight Test Institute
	NASTAR
	National Test Pilot School
	Space Adventures
	Starfighters Aerospace
	Waypoint 2 Space
	Zero G Corporation
Spaceports	Cecil Field
	Mid-Atlantic Regional Spaceport
	Oklahoma Air and Space Port
	Space Florida
	Titusville-Cocoa Airport Authority
Industry Organizations	Aerospace Industries Association
	Commercial Spaceflight Federation
	National Association of Spaceports
	The Tauri Group

Industry category	Organization/individual interviewed
Individuals	Dr. Frans von der Dunk, Harvey & Susan Perlman Alumni and Othmer Professor of Space Law, Nebraska College of Law, University of Nebraska-Lincoln
	Dr. Henry Hertzfeld, Research Professor of Space Policy and International Affairs, Space Policy Institute, George Washington University
	Dr. Justin Karl, Assistant Professor, College of Aviation, Embry-Riddle Aeronautical University
	Michael Lopez-Alegria, Principal, MLA Space
	Oscar Garcia, Chairman & CEO, InterFlight Global Corporation
	Dr. Ruth Stilwell, Adjunct Professor, Norwich University

Source: GAO. | GAO-17-100.

Appendix II: GAO Contacts and Staff Acknowledgments

GAO Contact	Gerald L. Dillingham, Ph.D., (202) 512-2834, or dillinghamg@gao.gov
Staff Acknowledgments	In addition to the individual named above, Cathy Colwell (Assistant Director), Stephanie Purcell (Analyst in Charge), Namita Bhatia-Sabharwal, Dave Hooper, Sara Ann Moessbauer, Amy Rosewarne, and Travis Schwartz made key contributions to this report.

GAO's Mission	The Government Accountability Office, the audit, evaluation, and investigative arm of Congress, exists to support Congress in meeting its constitutional responsibilities and to help improve the performance and accountability of the federal government for the American people. GAO examines the use of public funds; evaluates federal programs and policies; and provides analyses, recommendations, and other assistance to help Congress make informed oversight, policy, and funding decisions. GAO's commitment to good government is reflected in its core values of accountability, integrity, and reliability.
Obtaining Copies of GAO Reports and Testimony	The fastest and easiest way to obtain copies of GAO documents at no cost is through GAO's website (http://www.gao.gov). Each weekday afternoon, GAO posts on its website newly released reports, testimony, and correspondence. To have GAO e-mail you a list of newly posted products, go to http://www.gao.gov and select "E-mail Updates."
Order by Phone	The price of each GAO publication reflects GAO's actual cost of production and distribution and depends on the number of pages in the publication and whether the publication is printed in color or black and white. Pricing and ordering information is posted on GAO's website, http://www.gao.gov/ordering.htm. Place orders by calling (202) 512-6000, toll free (866) 801-7077, or TDD (202) 512-2537. Orders may be paid for using American Express, Discover Card, MasterCard, Visa, check, or money order. Call for additional information.
Connect with GAO	Connect with GAO on Facebook, Flickr, Twitter, and YouTube. Subscribe to our RSS Feeds or E-mail Updates. Listen to our Podcasts. Visit GAO on the web at www.gao.gov.
To Report Fraud, Waste, and Abuse in Federal Programs	Contact: Website: http://www.gao.gov/fraudnet/fraudnet.htm E-mail: fraudnet@gao.gov Automated answering system: (800) 424-5454 or (202) 512-7470
Congressional Relations	Katherine Siggerud, Managing Director, siggerudk@gao.gov, (202) 512-4400, U.S. Government Accountability Office, 441 G Street NW, Room 7125, Washington, DC 20548
Public Affairs	Chuck Young, Managing Director, youngc1@gao.gov, (202) 512-4800 U.S. Government Accountability Office, 441 G Street NW, Room 7149 Washington, DC 20548
Strategic Planning and External Liaison	James-Christian Blockwood, Managing Director, spel@gao.gov, (202) 512-4707 U.S. Government Accountability Office, 441 G Street NW, Room 7814, Washington, DC 20548

Please Print on Recycled Paper.